Student Lab Notebook

Long Version

BROOKS/COLE
CENGAGE Learning™

Australia • Brazil • Japan • Korea • Mexico • Singapore • Spain • United Kingdom • United States

Emergency Information

In case of emergency on campus, call..

In case of accident, please notify:

Name .... Relationship ....

Phone number(s) ....

BROOKS/COLE
CENGAGE Learning™

Student Lab Notebook, Long Version

Sr. Production Manager: Charlene Squibb
Developmental Editor: Ed Dodd
Cover & Text Designer: Gabriel Sim-Laramee

Long Version
ISBN-13: 978-0-03-027288-2
ISBN-10: 0-03-027288-2

Short Version
ISBN-13: 978-0-03-027289-9
ISBN-10: 0-03-027289-0

Brooks/Cole
10 Davis Drive
Belmont, CA 94002-3098
USA

Cengage Learning is a leading provider of customized learning solutions with office locations around the globe, including Singapore, the United Kingdom, Australia, Mexico, Brazil, and Japan. Locate your local office at:
international.cengage.com/region

Cengage Learning products are represented in Canada by Nelson Education, Ltd.

For your course and learning solutions, visit **academic.cengage.com**

Purchase any of our products at your local college store or at our preferred online store **www.ichapters.com**

Printed in the United States of America
9 10 11 11 10

Table of Contents

Table of Contents

Date	Experiment Title		Expt. No.

Name	Group No.	Grade

Signature

Note: Before writing on graph paper, place back flap under yellow sheet.

480200 CHG 7/21

Date	Experiment Title		Expt. No.

Name	Group No.	Grade

01

Signature

Note: Before writing on graph paper, place back flap under yellow sheet.

Date	Experiment Title		Expt. No.
Name	Group No.	Grade	

Signature

Note: Before writing on graph paper, place back flap under yellow sheet.

Date	Experiment Title		Expt. No.
Name	Group No.	Grade	

Signature

Note: Before writing on graph paper, place back flap under yellow sheet.

Date	Experiment Title		Expt. No.
Name	Group No.	Grade	

03

Signature

Note: Before writing on graph paper, place back flap under yellow sheet.

Date	Experiment Title		Expt. No.

Name	Group No.	Grade

Signature

Note: Before writing on graph paper, place back flap under yellow sheet.

480200 CHG 7/21

Date	Experiment Title		Expt. No.

Name	Group No.	Grade

Signature

Note: Before writing on graph paper, place back flap under yellow sheet.

Date	Experiment Title		Expt. No.

Name	Group No.	Grade

Signature

Date	Experiment Title		Expt. No.

Name	Group No.	Grade

Signature

Note: Before writing on graph paper, place back flap under yellow sheet.

Date	Experiment Title		Expt. No.

Name	Group No.	Grade

Signature

Note: Before writing on graph paper, place back flap under yellow sheet.

480200 CHG 7/21

Date	Experiment Title		Expt. No.

Name	Group No.	Grade

06

Signature

Note: Before writing on graph paper, place back flap under yellow sheet.

Date	Experiment Title		Expt. No.

Name	Group No.	Grade

Signature

Note: Before writing on graph paper, place back flap under yellow sheet.

480200 CHG 7/21

Date	Experiment Title		Expt. No.

Name	Group No.	Grade

Signature

Note: Before writing on graph paper, place back flap under yellow sheet.

Date	Experiment Title		Expt. No.

Name	Group No.	Grade

Signature

Note: Before writing on graph paper, place back flap under yellow sheet.

480200 CHG 7/21

Date	Experiment Title		Expt. No.

Name	Group No.	Grade

Signature

Note: Before writing on graph paper, place back flap under yellow sheet.

480200 CHG 7/21

Date	Experiment Title		Expt. No.
Name		Group No.	Grade

Signature

Note: Before writing on graph paper, place back flap under yellow sheet.

Date	Experiment Title		Expt. No.
Name	Group No.	Grade	

Signature

Note: Before writing on graph paper, place back flap under yellow sheet.

Date	Experiment Title		Expt. No.

Name	Group No.	Grade

Signature

Note: Before writing on graph paper, place back flap under yellow sheet.

480200 CHG 7/21

Date	Experiment Title		Expt. No.
Name	Group No.	Grade	

Signature

Note: Before writing on graph paper, place back flap under yellow sheet.

Date	Experiment Title		Expt. No.

Name	Group No.	Grade

Signature

Note: Before writing on graph paper, place back flap under yellow sheet.

480200 CHG 7/21

Date	Experiment Title		Expt. No.

Name	Group No.	Grade

Signature

Note: Before writing on graph paper, place back flap under yellow sheet.

480200 CHG 7/21

Date	Experiment Title		Expt. No.

Name	Group No.	Grade

Signature

Note: Before writing on graph paper, place back flap under yellow sheet.

480200 CHG 7/21

Date	Experiment Title		Expt. No.

Name	Group No.	Grade

Signature

Note: Before writing on graph paper, place back flap under yellow sheet.

Date	Experiment Title		Expt. No.
Name		Group No.	Grade

Signature

Note: Before writing on graph paper, place back flap under yellow sheet.

Date	Experiment Title		Expt. No.

Name	Group No.	Grade

Signature

Note: Before writing on graph paper, place back flap under yellow sheet.

480200 CHG 7/21

Date	Experiment Title		Expt. No.

Name	Group No.	Grade

Signature

Note: Before writing on graph paper, place back flap under yellow sheet.

480200 CHG 7/21

Date	Experiment Title		Expt. No.

Name	Group No.	Grade

Signature

Note: Before writing on graph paper, place back flap under yellow sheet.

480200 CHG 7/21

Date	Experiment Title		Expt. No.

Name	Group No.	Grade

Signature

Note: Before writing on graph paper, place back flap under yellow sheet.

Date	Experiment Title		Expt. No.

Name	Group No.	Grade

Signature

Note: Before writing on graph paper, place back flap under yellow sheet.

480200 CHG 7/21

Date	Experiment Title		Expt. No.

Name	Group No.	Grade

Signature

Note: Before writing on graph paper, place back flap under yellow sheet.

480200 CHG 7/21

Date	Experiment Title		Expt. No.

Name	Group No.	Grade

Signature

Note: Before writing on graph paper, place back flap under yellow sheet.

480200 CHG 7/21

Date	Experiment Title		Expt. No.

Name	Group No.	Grade

Signature

Note: Before writing on graph paper, place back flap under yellow sheet.

480200 CHG 7/21

Date	Experiment Title		Expt. No.
Name	Group No.	Grade	

Signature

Note: Before writing on graph paper, place back flap under yellow sheet.

480200 CHG 7/21

Date	Experiment Title		Expt. No.
Name	Group No.	Grade	

Signature

Note: Before writing on graph paper, place back flap under yellow sheet.

480200 CHG 7/21

Date	Experiment Title		Expt. No.
Name	Group No.	Grade	

Signature

Note: Before writing on graph paper, place back flap under yellow sheet.

480200 CHG 7/21

Date	Experiment Title		Expt. No.
Name		Group No.	Grade

Signature

Note: Before writing on graph paper, place back flap under yellow sheet.

Date	Experiment Title		Expt. No.

Name	Group No.	Grade

Signature

Note: Before writing on graph paper, place back flap under yellow sheet.

Date	Experiment Title		Expt. No.
Name	Group No.	Grade	

Signature

Date	Experiment Title		Expt. No.

Name	Group No.	Grade

Signature

Note: Before writing on graph paper, place back flap under yellow sheet.

Date	Experiment Title		Expt. No.

Name	Group No.	Grade

Signature

Note: Before writing on graph paper, place back flap under yellow sheet.

480200 CHG 7/21

Date	Experiment Title		Expt. No.

Name	Group No.	Grade

Signature

Note: Before writing on graph paper, place back flap under yellow sheet.

Date	Experiment Title		Expt. No.

Name	Group No.	Grade

Signature

Note: Before writing on graph paper, place back flap under yellow sheet.

Date	Experiment Title		Expt. No.
Name		Group No.	Grade

Signature

Note: Before writing on graph paper, place back flap under yellow sheet.

480200 CHG 7/21

Date	Experiment Title		Expt. No.

Name	Group No.	Grade

Signature

Note: Before writing on graph paper, place back flap under yellow sheet.

480200 CHG 7/21

Date	Experiment Title		Expt. No.

Name	Group No.	Grade

Signature

Date	Experiment Title		Expt. No.

Name	Group No.	Grade

Signature

480200 CHG 7/21

Date	Experiment Title		Expt. No.

Name	Group No.	Grade

Signature

Note: Before writing on graph paper, place back flap under yellow sheet.

480200 CHG 7/21

Date	Experiment Title		Expt. No.
Name		**Group No.**	**Grade**

Signature

Note: Before writing on graph paper, place back flap under yellow sheet.

480200 CHG 7/21

Date	Experiment Title		Expt. No.
Name	Group No.	Grade	

Signature

Note: Before writing on graph paper, place back flap under yellow sheet.

480200 CHG 7/21

Date	Experiment Title		Expt. No.

Name	Group No.	Grade

Signature

Note: Before writing on graph paper, place back flap under yellow sheet.

Date	Experiment Title		Expt. No.

Name	Group No.	Grade

Signature

Note: Before writing on graph paper, place back flap under yellow sheet.

480200 CHG 7/21

Date	Experiment Title		Expt. No.
Name	Group No.	Grade	

Signature

Note: Before writing on graph paper, place back flap under yellow sheet.

480200 CHG 7/21

Date	Experiment Title		Expt. No.

Name	Group No.	Grade

Signature

Note: Before writing on graph paper, place back flap under yellow sheet.

480200 CHG 7/21

Date	Experiment Title		Expt. No.

Name	Group No.	Grade

Signature

Note: Before writing on graph paper, place back flap under yellow sheet.

Date	Experiment Title		Expt. No.

Name	Group No.	Grade

Signature

Note: Before writing on graph paper, place back flap under yellow sheet.

480200 CHG 7/21

Date	Experiment Title		Expt. No.

Name	Group No.	Grade

Signature

Note: Before writing on graph paper, place back flap under yellow sheet.

480200 CHG 7/21

Date	Experiment Title		Expt. No.
Name	Group No.	Grade	

Signature

Note: Before writing on graph paper, place back flap under yellow sheet.

480200 CHG 7/21

Date	Experiment Title		Expt. No.

Name	Group No.	Grade

Signature

Note: Before writing on graph paper, place back flap under yellow sheet.

480200 CHG 7/21

Date	Experiment Title		Expt. No.
Name	Group No.	Grade	

Signature

Note: Before writing on graph paper, place back flap under yellow sheet.

480200 CHG 7/21

Date	Experiment Title		Expt. No.
Name	Group No.	Grade	

Signature

Note: Before writing on graph paper, place back flap under yellow sheet.

Date	Experiment Title		Expt. No.

Name	Group No.	Grade

Signature

Note: Before writing on graph paper, place back flap under yellow sheet.

Date	Experiment Title		Expt. No.

Name	Group No.	Grade

Signature

Date	Experiment Title		Expt. No.

Name	Group No.	Grade

Signature

Note: Before writing on graph paper, place back flap under yellow sheet.

480200 CHG 7/21

Date	Experiment Title		Expt. No.

Name	Group No.	Grade

Signature

Note: Before writing on graph paper, place back flap under yellow sheet.

Date	Experiment Title		Expt. No.
Name	Group No.	Grade	

Signature

Note: Before writing on graph paper, place back flap under yellow sheet.

480200 CHG 7/21

Date	Experiment Title		Expt. No.

Name	Group No.	Grade

Signature

Note: Before writing on graph paper, place back flap under yellow sheet.

480200 CHG 7/21

Date	Experiment Title		Expt. No.

Name	Group No.	Grade

Signature

Note: Before writing on graph paper, place back flap under yellow sheet.

480200 CHG 7/21

Date	Experiment Title		Expt. No.

Name	Group No.	Grade

Signature

Note: Before writing on graph paper, place back flap under yellow sheet.

480200 CHG 7/21

Date	Experiment Title		Expt. No.

Name	Group No.	Grade

Signature

Note: Before writing on graph paper, place back flap under yellow sheet.

480200 CHG 7/21

Date	Experiment Title		Expt. No.
Name	Group No.	Grade	

Signature

Note: Before writing on graph paper, place back flap under yellow sheet.

Date	Experiment Title		Expt. No.

Name	Group No.	Grade

Signature

Note: Before writing on graph paper, place back flap under yellow sheet.

480200 CHG 7/21

Date	Experiment Title		Expt. No.

Name	Group No.	Grade

Signature

Note: Before writing on graph paper, place back flap under yellow sheet.

Date	Experiment Title		Expt. No.

Name	Group No.	Grade

Signature

Note: Before writing on graph paper, place back flap under yellow sheet.

480200 CHG 7/21

Date	Experiment Title		Expt. No.
Name	Group No.	Grade	

Signature

Note: Before writing on graph paper, place back flap under yellow sheet.

480200 CHG 7/21

Date	Experiment Title		Expt. No.

Name	Group No.	Grade

Signature

Note: Before writing on graph paper, place back flap under yellow sheet.

Date	Experiment Title		Expt. No.

Name	Group No.	Grade

Signature

Note: Before writing on graph paper, place back flap under yellow sheet.

480200 CHG 7/21

Date	Experiment Title		Expt. No.
Name	Group No.	Grade	

Signature

Note: Before writing on graph paper, place back flap under yellow sheet.

Date	Experiment Title		Expt. No.

Name	Group No.	Grade

Signature

Note: Before writing on graph paper, place back flap under yellow sheet.

480200 CHG 7/21

Date	Experiment Title		Expt. No.

Name	Group No.	Grade

Signature

Note: Before writing on graph paper, place back flap under yellow sheet.

480200 CHG 7/21

Date	Experiment Title		Expt. No.
			Signature

Name	Group No.	Grade

Signature

Note: Before writing on graph paper, place back flap under yellow sheet.

Date	Experiment Title		Expt. No.

Name	Group No.	Grade

Signature

Date	Experiment Title		Expt. No.

Name	Group No.	Grade

Signature

Date	Experiment Title		Expt. No.

Name	Group No.	Grade

Signature

Date	Experiment Title		Expt. No.

Name	Group No.	Grade

Signature

Note: Before writing on graph paper, place back flap under yellow sheet.

480200 CHG 7/21

Date	Experiment Title		Expt. No.

Name	Group No.	Grade

Signature

Note: Before writing on graph paper, place back flap under yellow sheet.

Date	Experiment Title		Expt. No.

Name	Group No.	Grade

Signature

Note: Before writing on graph paper, place back flap under yellow sheet.

Date	Experiment Title		Expt. No.

Name	Group No.	Grade

Signature

Note: Before writing on graph paper, place back flap under yellow sheet.

44

Date	Experiment Title		Expt. No.

Name	Group No.	Grade

Signature

Note: Before writing on graph paper, place back flap under yellow sheet.

Date	Experiment Title		Expt. No.

Name	Group No.	Grade

Signature

Note: Before writing on graph paper, place back flap under yellow sheet.

480200 CHG 7/21

Date	Experiment Title		Expt. No.

Name	Group No.	Grade

Signature

Note: Before writing on graph paper, place back flap under yellow sheet.

Date	Experiment Title		Expt. No.

Name	Group No.	Grade

Signature

Note: Before writing on graph paper, place back flap under yellow sheet.

Date	Experiment Title		Expt. No.

Name	Group No.	Grade

Signature

Note: Before writing on graph paper, place back flap under yellow sheet.

480200 CHG 7/21

Date	Experiment Title		Expt. No.

Name	Group No.	Grade

Signature

Note: Before writing on graph paper, place back flap under yellow sheet.

Date	Experiment Title		Expt. No.
Name	Group No.	Grade	

Signature

Note: Before writing on graph paper, place back flap under yellow sheet.

Date	Experiment Title		Expt. No.
Name		Group No.	Grade

Signature

Note: Before writing on graph paper, place back flap under yellow sheet.

480200 CHG 7/21

Date	Experiment Title		Expt. No.

Name	Group No.	Grade

Signature

Note: Before writing on graph paper, place back flap under yellow sheet.

Date	Experiment Title		Expt. No.

Name	Group No.	Grade

Signature

Note: Before writing on graph paper, place back flap under yellow sheet.

480200 CHG 7/21

Date	Experiment Title		Expt. No.

Name	Group No.	Grade

Signature

Note: Before writing on graph paper, place back flap under yellow sheet.

480200 CHG 7/21

Date	Experiment Title		Expt. No.

Name	Group No.	Grade

Signature

Note: Before writing on graph paper, place back flap under yellow sheet.

Date	Experiment Title		Expt. No.

Name	Group No.	Grade

Signature

Note: Before writing on graph paper, place back flap under yellow sheet.

480200 CHG 7/21

Date	Experiment Title		Expt. No.

Name	Group No.	Grade

Signature

Note: Before writing on graph paper, place back flap under yellow sheet.

Date	Experiment Title		Expt. No.

Name	Group No.	Grade

Signature

Note: Before writing on graph paper, place back flap under yellow sheet.

480200 CHG 7/21

Date	Experiment Title		Expt. No.
Name	Group No.	Grade	

Signature

Note: Before writing on graph paper, place back flap under yellow sheet.

480200 CHG 7/21

Date	Experiment Title		Expt. No.
Name		Group No.	Grade

Signature

Note: Before writing on graph paper, place back flap under yellow sheet.

480200 CHG 7/21

Date	Experiment Title		Expt. No.

Name	Group No.	Grade

Signature

Note: Before writing on graph paper, place back flap under yellow sheet.

480200 CHG 7/21

Date	Experiment Title		Expt. No.

Name	Group No.	Grade

Signature

Note: Before writing on graph paper, place back flap under yellow sheet.

480200 CHG 7/21

Date	Experiment Title		Expt. No.

Name	Group No.	Grade

Signature

Note: Before writing on graph paper, place back flap under yellow sheet.

480200 CHG 7/21

Date	Experiment Title		Expt. No.
Name	Group No.	Grade	

Signature

Note: Before writing on graph paper, place back flap under yellow sheet.

Date	Experiment Title		Expt. No.

Name	Group No.	Grade

Signature

Note: Before writing on graph paper, place back flap under yellow sheet.

Date	Experiment Title		Expt. No.

Name	Group No.	Grade

Signature

Date	Experiment Title		Expt. No.

Name	Group No.	Grade

Signature

480200 CHG 7/21

Date	Experiment Title		Expt. No.
Name		Group No.	Grade

Signature

Note: Before writing on graph paper, place back flap under yellow sheet.

480200 CHG 7/21

Date	Experiment Title		Expt. No.

Name	Group No.	Grade

Signature

Note: Before writing on graph paper, place back flap under yellow sheet.

Date	Experiment Title		Expt. No.
Name		Group No.	Grade

Signature

480200 CHG 7/21

Date	Experiment Title		Expt. No.

Name	Group No.	Grade

Signature

Note: Before writing on graph paper, place back flap under yellow sheet.

Date	Experiment Title		Expt. No.
Name	Group No.	Grade	

Signature

Note: Before writing on graph paper, place back flap under yellow sheet.

Date	Experiment Title		Expt. No.

Name	Group No.	Grade

Signature

Note: Before writing on graph paper, place back flap under yellow sheet.

480200 CHG 7/21

Date	Experiment Title		Expt. No.

Name	Group No.	Grade

Signature

Note: Before writing on graph paper, place back flap under yellow sheet.

Date	Experiment Title		Expt. No.

Name	Group No.	Grade

Signature

Note: Before writing on graph paper, place back flap under yellow sheet.

Date	Experiment Title		Expt. No.

Name	Group No.	Grade

Signature

Note: Before writing on graph paper, place back flap under yellow sheet.

Date	Experiment Title		Expt. No.

Name	Group No.	Grade

Signature

480200 CHG 7/21

Date	Experiment Title		Expt. No.

Name	Group No.	Grade

Signature

Note: Before writing on graph paper, place back flap under yellow sheet.

Date	Experiment Title		Expt. No.

Name	Group No.	Grade

Signature

Note: Before writing on graph paper, place back flap under yellow sheet.

Date	Experiment Title		Expt. No.

Name	Group No.	Grade

Signature

Note: Before writing on graph paper, place back flap under yellow sheet.

Date	Experiment Title		Expt. No.
Name	Group No.	Grade	

Signature

Note: Before writing on graph paper, place back flap under yellow sheet.

480200 CHG 7/21

Date	Experiment Title		Expt. No.
Name	Group No.	Grade	

Signature

Note: Before writing on graph paper, place back flap under yellow sheet.

480200 CHG 7/21

Date	Experiment Title		Expt. No.

Name	Group No.	Grade

Signature

Note: Before writing on graph paper, place back flap under yellow sheet.

Date	Experiment Title		Expt. No.

Name	Group No.	Grade

Signature

480200 CHG 7/21

Date	Experiment Title		Expt. No.

Name	Group No.	Grade

Signature

Note: Before writing on graph paper, place back flap under yellow sheet.

480200 CHG 7/21

Date	Experiment Title		Expt. No.

Name	Group No.	Grade

Signature

Note: Before writing on graph paper, place back flap under yellow sheet.

Date	Experiment Title		Expt. No.

Name	Group No.	Grade

Signature

Note: Before writing on graph paper, place back flap under yellow sheet.

480200 CHG 7/21

Date	Experiment Title		Expt. No.

Name	Group No.	Grade

Signature

Note: Before writing on graph paper, place back flap under yellow sheet.

480200 CHG 7/21

Date	Experiment Title		Expt. No.

Name	Group No.	Grade

Signature

Note: Before writing on graph paper, place back flap under yellow sheet.

480200 CHG 7/21

Date	Experiment Title		Expt. No.
Name	Group No.	Grade	

Signature

Note: Before writing on graph paper, place back flap under yellow sheet.

480200 CHG 7/21

Date	Experiment Title		Expt. No.

Name	Group No.	Grade

Signature

Note: Before writing on graph paper, place back flap under yellow sheet.

83

480200 CHG 7/21

Date	Experiment Title		Expt. No.

Name	Group No.	Grade

Signature

Note: Before writing on graph paper, place back flap under yellow sheet.

480200 CHG 7/21

Date	Experiment Title		Expt. No.

Name	Group No.	Grade

Signature

Note: Before writing on graph paper, place back flap under yellow sheet.

480200 CHG 7/21

Date	Experiment Title		Expt. No.

Name	Group No.	Grade

Signature

Note: Before writing on graph paper, place back flap under yellow sheet.

Date	Experiment Title		Expt. No.
Name	Group No.	Grade	

Signature

Note: Before writing on graph paper, place back flap under yellow sheet.

480200 CHG 7/21

Date	Experiment Title		Expt. No.
Name		Group No.	Grade

Signature

Note: Before writing on graph paper, place back flap under yellow sheet.

Date	Experiment Title		Expt. No.

Name	Group No.	Grade

Signature

Date	Experiment Title		Expt. No.

Name	Group No.	Grade

Signature

Note: Before writing on graph paper, place back flap under yellow sheet.

480200 CHG 7/21

Date	Experiment Title		Expt. No.
Name	Group No.	Grade	

Signature

Note: Before writing on graph paper, place back flap under yellow sheet.

Date	Experiment Title		Expt. No.

Name	Group No.	Grade

Signature

Note: Before writing on graph paper, place back flap under yellow sheet.

480200 CHG 7/21

Date	Experiment Title		Expt. No.

Name	Group No.	Grade

Signature

Note: Before writing on graph paper, place back flap under yellow sheet.

Date	Experiment Title		Expt. No.
Name	Group No.	Grade	

Signature

480200 CHG 7/21

Date	Experiment Title		Expt. No.
Name	Group No.	Grade	

Signature

Note: Before writing on graph paper, place back flap under yellow sheet.

480200 CHG 7/21

Date	Experiment Title		Expt. No.
Name	Group No.	Grade	

Signature

Note: Before writing on graph paper, place back flap under yellow sheet.

c.)

480200 CHG 7/21

Date	Experiment Title		Expt. No.

Name	Group No.	Grade

Signature

Note: Before writing on graph paper, place back flap under yellow sheet.

Date	Experiment Title		Expt. No.
Name		Group No.	Grade

Signature

Note: Before writing on graph paper, place back flap under yellow sheet.

Date	Experiment Title		Expt. No.
Name		Group No.	Grade

Signature

Note: Before writing on graph paper, place back flap under yellow sheet.

480200 CHG 7/21

Date	Experiment Title		Expt. No.

Name	Group No.	Grade

Signature

Note: Before writing on graph paper, place back flap under yellow sheet.

Date	Experiment Title		Expt. No.

Name	Group No.	Grade

Signature

480200 CHG 7/21

Date	Experiment Title		Expt. No.

Name	Group No.	Grade

Signature

Note: Before writing on graph paper, place back flap under yellow sheet.

480200 CHG 7/21

Date	Experiment Title		Expt. No.

Name	Group No.	Grade

Signature

Note: Before writing on graph paper, place back flap under yellow sheet.

Date	Experiment Title		Expt. No.

Name	Group No.	Grade

Signature

Note: Before writing on graph paper, place back flap under yellow sheet.

480200 CHG 7/21

Date	Experiment Title		Expt. No.

Name	Group No.	Grade

Signature

Note: Before writing on graph paper, place back flap under yellow sheet.

Date	Experiment Title		Expt. No.

Name	Group No.	Grade

Signature

Note: Before writing on graph paper, place back flap under yellow sheet.

Date	Experiment Title		Expt. No.
Name		Group No.	Grade

Signature

Note: Before writing on graph paper, place back flap under yellow sheet.

08

480200 CHG 7/21

Date	Experiment Title		Expt. No.
Name	Group No.	Grade	

Signature

Note: Before writing on graph paper, place back flap under yellow sheet.

Date	Experiment Title		Expt. No.

Name	Group No.	Grade

Signature

Note: Before writing on graph paper, place back flap under yellow sheet.

Date	Experiment Title		Expt. No.

Name	Group No.	Grade

Signature

Date	Experiment Title		Expt. No.
Name	Group No.	Grade	

Signature

Note: Before writing on graph paper, place back flap under yellow sheet.

480200 CHG 7/21

Date	Experiment Title		Expt. No.
Name	Group No.	Grade	

Signature

Note: Before writing on graph paper, place back flap under yellow sheet.

480200 CHG 7/21

Date	Experiment Title		Expt. No.
Name	Group No.	Grade	

Signature

Note: Before writing on graph paper, place back flap under yellow sheet.

Date	Experiment Title		Expt. No.

Name	Group No.	Grade

Signature

Note: Before writing on graph paper, place back flap under yellow sheet.

A3

480200 CHG 7/21

Date	Experiment Title		Expt. No.

Name	Group No.	Grade

Signature

Note: Before writing on graph paper, place back flap under yellow sheet.

A8

Date	Experiment Title		Expt. No.

Name	Group No.	Grade

Signature

Note: Before writing on graph paper, place back flap under yellow sheet.

Date	Experiment Title		Expt. No.
Name	Group No.	Grade	

Signature

Note: Before writing on graph paper, place back flap under yellow sheet.

480200 CHG 7/21

Date	Experiment Title		Expt. No.

Name	Group No.	Grade

Signature

Note: Before writing on graph paper, place back flap under yellow sheet.

Date	Experiment Title		Expt. No.
Name		Group No.	Grade

Signature

Note: Before writing on graph paper, place back flap under yellow sheet.

Date	Experiment Title		Expt. No.

Name	Group No.	Grade

Signature

Note: Before writing on graph paper, place back flap under yellow sheet.

480200 CHG 7/21

Date	Experiment Title		Expt. No.

Name	Group No.	Grade

Signature

Note: Before writing on graph paper, place back flap under yellow sheet.

Date	Experiment Title		Expt. No.

Name	Group No.	Grade

Signature

Note: Before writing on graph paper, place back flap under yellow sheet.

Date	Experiment Title		Expt. No.

Name	Group No.	Grade

Signature

Note: Before writing on graph paper, place back flap under yellow sheet.

Date	Experiment Title		Expt. No.

Name	Group No.	Grade

Signature

Note: Before writing on graph paper, place back flap under yellow sheet.

Date	Experiment Title		Expt. No.
Name	Group No.	Grade	

Signature

Note: Before writing on graph paper, place back flap under yellow sheet.

Date	Experiment Title		Expt. No.

Name	Group No.	Grade

Signature

Note: Before writing on graph paper, place back flap under yellow sheet.

Date	Experiment Title		Expt. No.

Name	Group No.	Grade

Signature

Note: Before writing on graph paper, place back flap under yellow sheet.

480200 CHG 7/21

Date	Experiment Title		Expt. No.

Name	Group No.	Grade

Signature

Note: Before writing on graph paper, place back flap under yellow sheet.

Date	Experiment Title		Expt. No.
Name	Group No.	Grade	

Signature

480200 CHG 7/21

Date	Experiment Title		Expt. No.

Name	Group No.	Grade

Signature

Note: Before writing on graph paper, place back flap under yellow sheet.

480200 CHG 7/21

Date | Experiment Title | Expt. No.

Name | Group No. | Grade

Signature

Note: Before writing on graph paper, place back flap under yellow sheet.

480200 CHG 7/21

Date	Experiment Title		Expt. No.

Name	Group No.	Grade

Signature

Note: Before writing on graph paper, place back flap under yellow sheet.

480200 CHG 7/21

Date	Experiment Title		Expt. No.
			Signature

Name	Group No.	Grade

Signature

Note: Before writing on graph paper, place back flap under yellow sheet.

Date	Experiment Title		Expt. No.

Name	Group No.	Grade

Signature

Note: Before writing on graph paper, place back flap under yellow sheet.

Date	Experiment Title		Expt. No.

Name	Group No.	Grade

Signature

Date	Experiment Title		Expt. No.

Name	Group No.	Grade

Signature

480200 CHG 7/21

Date	Experiment Title		Expt. No.

Name	Group No.	Grade

Signature

Note: Before writing on graph paper, place back flap under yellow sheet.

480200 CHG 7/21

Date	Experiment Title		Expt. No.

Name	Group No.	Grade

Signature

Note: Before writing on graph paper, place back flap under yellow sheet.

480200 GHG 7/21

Date	Experiment Title		Expt. No.

Name	Group No.	Grade

Signature

Note: Before writing on graph paper, place back flap under yellow sheet.

Date	Experiment Title		Expt. No.

Name	Group No.	Grade

Signature

Note: Before writing on graph paper, place back flap under yellow sheet.

Date	Experiment Title		Expt. No.

Name	Group No.	Grade

Signature

480200 CHG 7/21

Date	Experiment Title		Expt. No.
Name		Group No.	Grade

Signature

Note: Before writing on graph paper, place back flap under yellow sheet.

480200 CHG 7/21

Date	Experiment Title		Expt. No.
Name	Group No.	Grade	

Signature

Note: Before writing on graph paper, place back flap under yellow sheet.

480200 CHG 7/21

Date	Experiment Title		Expt. No.

Name	Group No.	Grade

Signature

Note: Before writing on graph paper, place back flap under yellow sheet.

Date	Experiment Title		Expt. No.

Name	Group No.	Grade

Signature

Note: Before writing on graph paper, place back flap under yellow sheet.

Date	Experiment Title		Expt. No.

Name	Group No.	Grade

Signature

Note: Before writing on graph paper, place back flap under yellow sheet.